CASE STUDY CONTRACTOR PERFORMANCE AND CAPABILITY EVALUATION

FULBODH CHAUDHARY – CONSULTANT (PLANT PERFORMANCE MANAGEMENT)
B.TECH (IIT DHANBAD)/SIX SIGMA/ LEAD AUDITOR

- ORGANISATION
  - ✓ TATA STEEL
  - ✓ LAFARGE CEMENT
  - ✓ NEYVELLI LIGNITE CORPORATION
  - ✓ HOBELLS CO. (CUMMINS GROUP)
  - ✓ CHING YUAN CO. (INTERNATIONAL ASSIGNMENT)

- ASSIGNMENTS EXPERIENCE CATEGORY:-

  * **MAINTENANCE MANAGEMENT**
    INDUSTRY: - CEMENT + MINES + MANUFACTURING
    WORK SCOPE:-

    1. PlANNED MAINTENANCE

       a Preventive maintenance
          1   Running type
          11 Shutdown type
          111 Schedule type

       b Corrective maintenance
          1   Breakdown type
          11 Shutdown type

       c Predictive maintenance

    2 UNPLANNED MAINTENANACE
       Emergency maintenance
    3 Maintenance planning and scheduling
    4 Total productive maintenance
    5, Providing redundancy
    6. Increasing repair capability
    7. Improve system reliability
    8 Mean time failure determination
    9 Maintenance cost analysis and control
    10 Autonomous maintenance
    11 Computerized maintenance
    12 Equipment life cycle management
    13 Beyond economic repair concept and disposal
    14 Spares management (insurance, non insurance, VED)

CONTRACTOR CAPABILITY EVALUATION		
Name & Address of Vendor:	Name of the Auditor	Fulbodh chaudhary
M/s.	TPL EmpID	TPL91895
	SR Number	232020
	Date of Visit	18-Feb-2020
	Report No.	232020-18-Feb-2020
	Type of Service / Product	Contractor (civil-construction of road and civil structural work including RCC of STP)
Contractor's Representative interacted for evaluation	Name of the site referred:	
Name : Mr.	sector 40. Panipat. haryana. Pincode 132103	
Cell No :		
Telephone No : 0		
Email :	Project :	--

	Evaluation Criteria	Max Points	Scored Points
1	Experience	10	4
2	Tools & Plant	15	7.35
3	Human Resources	15	14.75
4	Quality Management System	15	6.95
5	Environment / Health/ Safety	16	12
6	Statutory Requirements	4	2.5
7	General/ Administration	5	4.6
8	Financial Strength	20	18
	Total Score	100	70.15

Note:
Scores for all criteria- shall be evaluated by TQS.

Feedback from other customers (Telephonic Enquiry/ Personal Visit/Feedback letter/Team Visit) : -

Checked customer feedback/site clearance report from following customer and found the same satisfactory 1 construction of boundary wall client name Eldeco Infrastructure and properties Ltd,Panipat,Haryana Job awarded date 1.4.2014 and completion date 25.7.2017 2 TDI ,Panipat ,Haryana-construction of road, contract awarded date -23.2.2017 and completion date 16.3.2018 3 TDI,Panipat,Haryana-construction of boundary wall, contract awarded date 1.12.2015 and completion date 30.3.2018

Special Remarks:	
*Contractor is having experience since 2006 and have completed more than 3 projects of client enlisted in major customer list * Visited subject site of contractor and found the same site of M/s Ansal Infrastructure and property private Ltd and Eldeco for construction of Road and STP. Refer to file current site for work order and jobs done. Site photographs attached. Talked to client project Incharge as given below and found their feedback satisfactory 1 Eldeco, AGM PROJECT Munish garg 9996975773 2 Ansal Infra GM project DP Dagar 9996111606	
Strengths: * Retained skilled and experienced employees * Have satisfactory feedback of customer from the completed site taken on cell phone * Office set up and infrastructure facility observed good. *Have good fleet of mechanical excavation machinery and concrete preparation .	
Opportunity for improvement: * no evidence found for wage register and HR Register etc * no evidence found vehicle maintenance checklist and staff. * No evidence found for vendor evaluation and rating as contractor supplies material also	
Above Contractor is approved / not approved for enlistment for type and range of services indicated above.	
Evaluated by: Name, Signature & Date	Auditor- TQ Services
Reviewed by: Name, Signature & Date	Dy. Head Certification- TQ Services

Basic	Vendor Name	M/s.
	Establishment Year	2006 as per Vat registration
	Type of Contractor	Civil
Financials	Total Income / Revenue (in INR Lakhs.) Last Financial Year	1234.14 L
	Total Assets (in INR Lakhs.) Last Financial Year	1130.29 L
	Total Equity (in INR Lakhs.) Last Financial Year	224.16 L
	Total Debt (in INR Lakhs.) Last Financial Year	906.13 L
Manpower	Direct staff - Skilled (in nos.)	15
	Direct staff - unSkilled (in nos.)	100
	Contract - Skilled Manpower (in nos.)	0
	Contract- Unskilled Manpower (in nos.)	50
Machinery	Major Machinery capacity (in MT)	25 T DUMPER
Order Booking / Present Load	Total No of On going Industrial Projects (in Nos.)	4
	Total Value of Bal. Scope of On	2

	Projects (in Crs.)	
	Total No projects completing in next 3 months (in nos.)	0
	Total Value of projects completing in next 3 months (in Crs.)	0
Safety	Last 3 year LTI (Lost Time Injury) overall No.of Incidents	0
	Last 3 years Fatality if Any in **Industrial project ****	0
Quality & Safety Certification	ISO	No
	OSHAS	No
Statutory Documents	Labour License & Work contract registration	Available
Major Customers	Names	1 Eldeco,Panipat ,Haryana 2 TDI,Panipat ,Haryana 3 Ansal properties and Infrastructure, Gurgaon, Haryana

CAPABILITY EVALUATION REPORT OF CONTRACTOR					
(Civil, Mechanical, Electrical & Instrumentation)					
Sl. No	EXPERIENCE (Max Score – 10)				
1	Evaluation Criteria	Guidance (check last 3 years Records)	Points	Scored Points	Audit Evidences- (Pl mention details in brief)
1.1	Customer base in various Indian States/regions/Countries	Customer Base in Single State / Region			
	No Customer Base in State / Region	(In Any State Or Region)	0	1	Haryana refer to client completion certificates and work orders
	Customer Base in Single State / Region	Any State/Region	1		
	Customer Base in two States / Regions	Any Two Or More States/Regions	2		
	Customer Base in more States / Regions/Countries	Any Three States/Regions/Countries	3		

	Work Experience for Similar type of jobs executed	Or More **Executed two similar type of job in last 2 years = 3**			
1.2	Not executed similar type of jobs = 0	Similar projects like PG, T&D, WWW, Urban Infrastructure, OGH, Metals & Minerals, etc....	0	3	executed 3 projects since 2017 refer to completion certificates for Eldeco and TDI
	Executed one similar type of job in last 2-3 years = 1		1		
	Executed one similar type of job in last 2 years = 2		2		
	Executed two similar type of job in last 2 years = 3		3		
	Experience of most Critical/Precise Jobs	**Not Executed Critical/Precise Jobs**			
1.3	Not Executed Critical/Precise Jobs	Critical/ Precise meaning boiler erection, commissioning, storage tanks, pressure vessels, etc	0	0	no evidence found
	Executed one Critical/Precise Job		0.5		
	Executed more than one Critical/Precise Job		1		

	Experience in Criticalities/Adverse environment <u>YES = 1</u>, No =0				
1.4	Experience in Resolving Local Issues at least a case	Local issues such as employment, state restrictions, conflict management, safety, Contingency, etc.... (records or references required)	No	0	no evidence found
1.5	Experience of Working in Remote/Unrest Areas /J & K/ North Eastern States/South African Countries	Andaman, Arunachal Pradesh, Jharkhand, Bihar, Chattishgarh, Assam, Etc	No	0	have experience in state of Haryana only
1.6	Experience of Green Field Project	Projects involving plantation, energy saving landscaping, water conservation, animal conservation, green buildings etc	No	0	no evidence found
		Overall Score		**4**	

	CAPABILITY EVALUATION REPORT OF CONTRACTOR				
	(Civil, Mechanical, Electrical & Instrumentation)				
2	**TOOLS & PLANTS (Max Score -15)**				
Evaluation Criteria	Guidance (check last 3 years Records)		Points	Scored Points	Audit Evidences- (Pl

					mention details in brief)
2.1	**Layouts and Measurement**				
	Distance Measuring Devices Count			1	
2.1.1	Available	Distance Measuring tools, Long Measuring tape, Laser Measuring tools, Laser Dimension Master. **0.5 mark for each of the above instruments**	2	0.5	Long measuring tapes
	Partially Available		1		
	Not Available		0		
	Levels			2	
2.1.2	Available	Digital Electronic levels, Angle finders, Torpedo levels, Specialty levels, Magnetic conduit level, Laser Plumb bob, Electronic water level, Tripods, Laser Mounts, Grade Rods **0.25 mark for each of the above instrument**	2	0.5	water levels
	Partially Available		1		
	Not Available		0		
Sub Total					

2.2	Equipment's & Tools		Mechanical		
MECHANICAL Contractors					
2.2	**MATERIAL HANDLING TOOLS**				
	Available	Cranes, Hydra, Own tools & tackles like Rigs, Winches, Hydraulic Jacks, Chain Pulley Blocks & Wire Ropes, Compressors, Special Purpose Fixtures, Strain Gages Crane and compressors 0.5 mark each. Rest 0.2 marks each	**_Crane & Compressor Count_**		
	Partially Available		**_Rest of the equipment count_**		
	Not Available				
	WELDING SETS		**_Enter Count_**		
	Available	Rectifier, Transformer, MIG Sets, TIG Welding/Stick Welding/Cast Iron Welding Any welding machine with	2		
	Partially Available		1		

Not Available	rectifier transformer 0.25 mark each	0			
GAS CUTTING SETS		*Select*			
Available	Gas Regulator, Cutting Torch	2	0		
Partially Available		1			
Not Available		0			
TESTING & MEASURING		*Enter Count*			
Available	Flow Meters, Gauges, Pneumatic Measuring Devices, Roundness Tester and Surface Finish Measuring Instrument- Each type of instrument 0.25 mark	1			
Partially Available		0.5			
Not Available		0			

PERSONAL PROTECTIVE CLOTHING / FR CLOTHING		*Enter Count*			
Available	Poly guard Coveralls, PPE Kits - Safety Glass, Hard Hats, Arch Flash Jacket, Ear Defenders, Respirators, Insulated Gloves, Bib Overalls, Eye & Face Protectors 0.1 mark for each type of safety kit	1			
Partially Available		0.5			
Not Available		0			
Sub Total					

2.2	Equipment's & Tools		Civil		
CIVIL Contractors					
2.2	**EXCAVATION MACHINERY**		*Enter Count*	3	
	Available	JCB, Excavators, Road Roller, Backhoe Loader **Each type allot 0.5 marks**	2	1.5	excavator, road roller, Paver
	Partially Available		1		
	Not Available		0		
	CONCRETE PREPARATION		*Enter Count*	5	
	Available	Concrete Mixers, Vibrators, Compactors, Hot Mixers, Water Tankers /Tanks, Concrete Pumps,	2	1	ready mix machine, compactor, hot mixer, water tanker, vibrator
	Partially Available		1		
	Not Available		0		

		Rotary Sand Sieving Machine, Turbo Grinders, Floor Strippers & Scalers, Floor Strippers & Scalars, Scabblers, Scarifiers **Each item marks 0.2**			
	SCAFFOLDINGS & SHUTTERING		_**Enter Count**_	2	
	Available	Shuttering Plates, Adjustable Jacks, Swivel Coupler, Steel Channels **Each item marks = 0.5**	2	1	shuttering plates and steel channels
	Partially Available		1		
	Not Available		0		
	TESTING & MEASURING		_**Enter Count**_	2	
	Available	Concrete Compression Testers, Profometer, Concrete Permeability Tester, Pull-Off Tester, Rebar Detector, Concrete Test Hammer **Each item marks= 0.2**	1	0.4	Concrete testing machines
	Partially Available		0.5		
	Not Available		0		

	PERSONAL PROTECTIVE CLOTHING / FR CLOTHING		_**Enter Count**_	4	
	Available	Polyguard Coveralls, PPE Kits - Safety Glass, Hard Hats, Arch Flash Jacket, Ear Defenders, Respirators, Insulated Gloves, Bib Overalls, Eye & Face Protectors **0.1 mark for each type of safety kit**	1	0.4	helmet, safety shoes, Safety jackets, safety belt
	Partially Available		0.5		
	Not Available		0		
Sub Total				4.3	

2.2	Equipment's & Tools		Electrical		
Electrical Contractors					
2.1	**WIRE PULLING TOOLS**		_**Enter Count**_		
	Available	Pulling Rods, Pullers, Fish Tapes, Adjustable & Flexible Airbags, Cable Dispensers, Wire Racks,	2		
	Partially Available		1		

	Not Available	Non Conductive Wire Snake, Laser Cable Stringing Tool Each item = 0.25	0		
	INSULATED TOOL KITS		***Enter Count***		
	Available	Torque Wrenches, Hand Drills, Concrete Drills, Hammers, Spanners, Ratcheting Box Wrenches, Pliers, Screw Drivers, Ratchets, Extensions & Adapters, Powered Cable Cutters Each item =0.2	2		
	Partially Available		1		
	Not Available		0		
	MEASURING TOOLS		***Enter Count***		
	Available	Voltmeter, Ammeter, Tachometer, Megger, Clamp on Meters, Meghometers, Analogue/Digital Clamp meters Each item= 0.25	2		
	Partially Available		1		
	Not Available		0		
	PERSONAL PROTECTIVE CLOTHING / FR CLOTHING		***Enter Count***		
	Available	Polyguard Coveralls, PPE Kits - Safety Glass, Hard Hats, Arch Flash Jacket, Ear Defenders, Respirators, Insulated Gloves, Bib Overalls, Eye & Face Protectors 0.1 mark for each type of safety kit	2		
	Partially Available		0.5		
	Not Available		0		
Sub Total					

2.3	Office with Communication & Administrative Facilities (Available =1, Not Available=0)				
2.3.1	Telephones and admin staff, reception		Available	1	Cell phone and office available. Director is responsible for administration
2.3.2	**Software**	Availability of latest Software like AutoCAD MS office etc, all products (excel, word, powerpoint will be treated as 1 count / 0.25 points), 0.75 for	1	0.25	MS office

		other softwares available **0.25 for each of the software's available as applicable**				
2.3.3	**Hardward & Security**	Availability of workstations ,Computers, CD Writers, Scanners, Printers, Plotters etc. with adequate information security & data transmittal system. **Each of the device will carry 0.2 marks**	4		0.8	computer,laptop,scanners and printers
Sub Total					2.05	
Overall Score						

CAPABILITY EVALUATION REPORT OF CONTRACTOR
(Civil, Mechanical, Electrical & Instrumentation)

Sl. No.	HUMAN RESOURCES (15)						
	Manpower				Guidance	Points	Scored Points	Audit Evidences- (Pl mention details in brief)

Sl. No.						Guidance	Points	Scored Points	Audit Evidences
3.1	**Human Resource**	Business Volume INR (CR)			Verify HR records 10 years' experience of skilled manpower if not degree or diploma to be considered	Select Business Volume			B.tech 1 no with 5 years of experience.Diploma -3 nos with 4 years of experience. supervisors Engineers -5 nos with 4- 7 years of experience no evidence found for any HR records. This figure is based on as observed at site
		0-10	10-100	>=100	0-10				
	Degree or Diploma or Experienced Minimum no.	2	4	6	**4**		4		
	Minimum Supporting Staff (Foreman, Surveyors, Draftsman, Supervisors, Office Staff)	4	8	12	5		4		

3.2	SUB CONTRACTOR RESOURCES			Civil	
	Labour Contractor	Please award 0.25 for one tie up & 0.50 for two or more tie-ups Please ensure minimum 2 tie-ups by verifying work orders/any evidence	1	0.25	Labour requirement is off sourced .refer to file job work order. WO issued by vendor to M/s. Landbert Infrasture LLP enclosed.
	Bending Contractor (reinforced Steel)	Contractual agreements /Internal memos/ challans / attendance / registers to be checked	2	0.5	no off sourcing is done for this job, Using own Manpower
	Excavation Agencies		02	0.5	no off sourcing is done for this job, Using own Manpower
	Concrete Mixers providers		2	0.5	no off sourcing is done for this job, Using own Manpower
	Flooring & Tiling Agencies		2	0.5	not applicable
	Plastering Agencies		2	0.5	no off sourcing is done for this job, Using own Manpower
	Shuttering Agencies		2	0.5	no off sourcing is done for this job, Using own Manpower
	Woodwork/Carpentry		2	0.5	not applicable
	Painting Agencies		2	0.5	no off sourcing is done for this job, Using own Manpower
	Plumbing & Sanitation Agencies		2	0.5	no off sourcing is done for this, Using own Manpower
	Tractors/Trailors/Dumpers Providers		2	0.5	no off sourcing is done for this job, Using own Manpower
	Surveying & Topographical Agencies		2	0.5	not applicable
	Drawings & Drafting Agencies		2	0.5	not applicable
	Piling Contractors		2	0.5	no off sourcing is

			done for this job, Using own Manpower
Scored Points		6.75	

SUB CONTRACTOR RESOURCES		Electrical	
ELECTRICAL & INSTALLATION Contractors			
Contractors for small excavation	Please award 0.5 point for one tie up & 1 for two or more tie ups Please ensure minimum 2 tie-ups by verifying work orders/any evidence Contractual agreements /Internal memos/ challans	0	
Contractors for Cable Laying		0	
Contractors for Control Panel Jobs		0	
Contractors for Electrifications		0	
Contractors for allied civil works		0	
Contractors for substation & transformer work		0	
Liasoning Agency		0	
Scored Points		0	

SUB CONTRACTOR RESOURCES		Mechanical	
MECHANICAL Contractors			
Labour Contractor	Please award 0.5 point for one tie up & 1 for two or more tie ups Please ensure minimum 2 tie-ups by verifying work orders / any evidence Contractual agreements Internal memos / challans	0	
Small Fabricators		0	
Material handling Agencies		0	
Drawing & Design Agencies		0	
Mechanical Testing Agencies		0	
Painting & finishing Agencies		0	
Radiography Testing		0	
Scored Points		0	

| | | | | Overall Score | 14.75 |

CAPABILITY EVALUATION REPORT OF CONTRACTOR
(Civil, Mechanical, Electrical & Instrumentation)

Sl. No	Evaluation Criteria	Guidance	Points	Scored Points	Audit Evidences- (Pl mention details in brief)
4.1	QUALITY MANAGEMENT SYSTEM CERTIFICATION(Max Score -15)				
4.1.1 The Quality System complies with ISO 9001					
4.1.1.1	Company possess Valid ISO 9001 Certification	Scope and validity date to be verified (Yes = 0.25)	No	0	no evidence found
4.1.1.2	Is a written Manual of Quality Procedures available & maintained?	Quality manual, procedures (mandatory 6 for ISO 9001-2008 only) and others, essential work instructions	No	0	no evidence found
	Updating and Maintenance	Whether documents are getting revised at least yearly once	No	0	no evidence found
4.1.1.3	Awareness of Quality Systems as per ISO	Asking the personnel at least three persons at various levels about Quality Policy and Implementation at their department Quality policy and objectives deployment to be verified	No	0	do not have any quality system
	Company is having QA/QC Department	If Company is having QA/QC Department	No	0	quality taken by client
4.1.1.4	If Company is having Qualified Inspectors/Professional Engineers for QA/QC Departments	Relevant to the scope of activities, Relevant Tech. Qualification shall be verified	No	0	quality taken by client
Sub Total				0	

4.1.2 Document Control

| 4.1.2.1 | **Master List of Documents viz., Drawings, Purchase Orders & amendments; if any** | | | Not Maintained | |

		data sheets, BOQ, Technical Specifications, applicable Codes etc.				
		Maintained Fully	Verify documents maintained as per master list and approved, For each type of documents above marks allotted = 0.1.	0.50	0	no evidence found
		Maintained Partially		0.25		
		Not Maintained		0.00		
4.1.2.2	Correlation of Document Revisions with Master List of Documents		Verify revision numbers on documents tallying with master list sample 5 numbers	No	0	no evidence found
4.1.2.3	Inspection / Testing reports are identified with correct drawing numbers and revisions (Verify latest reports with drawing no's & revisions		Verify 5 inspection reports	Yes	0.5	inspection taken care by client
4.1.2.4	Availability of Current revision drawings at Works Place / Shop floor		Verify shop floor drawings 5 numbers being used revision numbers tallying with master list. Obsolete documents maintained identified.	Yes	0.5	drawing provided by client
Sub Total					1	
4.1.3 Purchase Control						
4.1.3.1	Vendor Evaluation Process		Whether selection of Vendors is being done Based on evaluation of certain parameters	No	0	no evidence found
4.1.3.2	Availability of Approved vendors list / Data base		Verify master list of approved vendors / approved vendor data base and check sample 5 POs and see whether approved vendors used	No	0	no evidence found
4.1.3.3	Purchase Order Preparation Technical Specifications are specified in the Indent/Enquiry/PO check 5 sample POs, indents and enquiries		check 5 sample POs, indents and Specifications to include revision numbers and inspection methodology and	Yes	0.25	cheked po and found delivery terms with specification mentioned there in.

		quality system to be mentioned in PO			
	Product Delivery Schedule	Check 5 sample pos and verify whether delivery schedule mentioned	Yes	0.25	found delivery schedule mentioned in PO
Sub Total				**0.5**	

4.2		Process control, Inspection and Testing, Calibration of Instruments and Control of Nonconforming Product			
4.2.1	**Process control**				
4.2.1.1	**Methodology : Availability of information and characteristics of the Product and Production viz., Job Card System, Procedures or Method Statements, Work Instructions etc.,**		Maintained Fully		
	Maintained Fully	In addition control plans / process parameters information to be available, 0.50 for product parameters as per QAP and 0.50 for process parameters	1.00	1	QAP taken care by client
	Maintained Partially		0.50		
	Not Maintained		0.00		
4.2.1.2	Measuring Instruments -	Availability and use of Monitoring and Measuring equipment (like Pressure Gauges, Total Stations, Megger Instruments etc.,) max of 0.25	Yes	0.25	taken care by client
	Availability of Lab Facilities	Verify for lab facilities existence and whether complying with ISO 17025 "0" for unaccredited lab and "0.25" for accredited lab.	Unaccredited	0	no evidence found
4.2.1.3	**Product Identification & Traceability through its Cycle of Production, During - Raw Material, Fabrication, Assembly and Final Stage**				
	Proper storage of Raw material is maintained	1. Items stored in designated areas (Racks) 2. Part Location Identifiable through system	5	0.5	All found taken care under client supervision

		3. Proper Preservation of items 4. Handling of items in appropriate manner and using appropriate material handling equipments 5. Non Confirming items properly identified 0.1 mark for each stage			
	Product Identification Maintained	1. Verify part number/ product identification 2. Inspection tagging for raw material 3. In process (fabrication) 4. Assembly 5. Final stages 0.1 mark for each stage.	5	0.5	All found taken care under client supervision
	Traceability through in-process stages	1. Records 2. Inspection tagging for raw material 3. In process (fabrication) 4. Assembly 5. Final stages 0.1 mark for each stage	5	0.5	contractor found doing R.C.C work under client supervision. Hence all found complied
Sub Total				2.75	

4.2.2	Inspection / Testing	
4.2.2.1	**Incoming Material Control viz., Storage, Identification, Document control and Correlation with certificates and FIFO etc.,**	

Storage and Identification is found acceptable	Material stored in identified racks and with proper tags and material stored outside is properly tagged, handling with proper equipments, preservation is maintained, FIFO followed. see sample of 5	4	0.4	storage,identification handling and preservation found satisfactory. FIFO no system found

		0.1 mark for each of above			
	Document Control is available in Stores	Receipt register, issue register and stock register maintained. Scrap register and tool register. Formats current revision being used. see sample of 5 0.1 mark for each of above	0	0	no evidence found
	Correlation of Documents and Stored Material	Physical verification of stock sample 5 and compare with book stock see sample of 5 0.1 mark for each of above	0	0	no evidence found for stock record
4.2.2.2	**Stage Inspection and Final Inspection activities carried out**				
	All Stage Inspection	Verify QAP and inspection reports of 5 sample inspections (all samples to be OK)	5	0.25	inspection taken care by client
	All Final Inspection		5	0.25	inspection taken care by client

	Maintenance of Stage Inspection and Final Inspection Records viz., Acceptance / Clearance records, Handing over records etc., including Third Party Inspection Reports / Records				
4.2.2.3	All Stage Inspection Records	Record retention of all project records till date from start date. Inspection reports approval verify. Identify whether accept/ reject mentioned in report (5 samples)	Yes	0.5	inspection taken care by client
	All Final Inspection Records		Yes	0.5	inspection taken care by client
Sub Total				**1.9**	

4.2.3	**Calibration of Instrument**				
	Working status of Measuring Instruments / Testing Equipments, Maintenance of Calibration Records and correlation				
4.2.3.1	Found in working condition	Verify sample 5 instruments and carries 0.20 marks per sample Equipment is in working condition and moving freely and with no damages	2	0.4	CTM machine and proving ring
	Calibration records available		2	0.4	ctm, proving ring
Sub Total				**0.8**	

4.2.4	**Control of Nonconforming Product**				
4.2.4.1	Monitoring of changes in Specifications of Product Records of Design/Drawing/Site Changes maintained	Verify 5 samples and each change record carries 0.10 points	0	0	drawing taken care by client
4.2.4.2	Records regarding Repair / Re-works is maintained	Verify 5 samples and each change record carries 0.10 points Authorized person giving disposition, identification of nonconformity	0	0	repair and rework taken care by client
4.2.4.3	Monitoring of Non-conformity of Product	Check nonconformity records and disposition by authorized person, identification of nonconformity, scrap register . review records of nonconformity in Company. Check 5 Samples	0	0	taken care by client engineer at site
	Resolution of Customer Complaints	Customer complaint handling records, corrective actions taken by identifying root cause, final communication to client on closure. Check 5 samples and average time for closure.	0	0	customer complains are resolved at site itself under client supervision
Sub Total				0	
Overall Score				6.95	

	CAPABILITY EVALUATION REPORT OF CONTRACTOR				
	(Civil, Mechanical, Electrical & Instrumentation)				
Sl. No	Evaluation Criteria	Guidance	Points	Scored Points	Audit Evidences- (Pl mention details in brief)
5	Environment /Health/Safety (Max Score – 16)	All Available = 1.0; Partially Available = 0.5; Not Available = 0			
5.1	Company possess Valid ISO 14001 and/or OHSAS 18001 Certification		Not Available	0	no evidence found
5.2	Safety Policy /		All	1	safety taken care by

	Statement on Safety by Contractor's Management		Available		client, Follows client policy
5.3	Organogram / details of Safety Officers employed in previous projects	Please verify Qualifications	All Available	1	safety taken care by client
5.4	Safety Statistics for past 3 years for major sites or wherever Safety Officer is deployed		All Available	1	It is in Client scope
5.5	Training report for past 3 years towards Safety Training of manpower		All Available	1	training taken care by client
5.6	Hazard Identification & Risk Assessment (HIRA) or Job Safety Analysis (JSA) documents for major activities undertaken		All Available	1	taken care by client
5.7	Records of Safety Meetings for sites where Safety Officer is deployed		All Available	1	safety taken care by client
5.8	Details of First-aid facilities and First-aider employed in previous projects		All Available	1	It is in Client scope
5.9	Records of medical examination conducted for their workmen or medical camps details conducted		Not Available	0	Not available
5.1	Records of Height pass / gate pass for their workmen.		All Available	1	taken care by client
5.11	Records of Third party inspection reports for Tools & Plants	Enclose reports	All Available	1	checked and found registration certificates available and attached. Contractor found maintaining fitness certificates for equipments. refer to file PDF kETAN
5.12	Accident / incident		All	1	taken care by client

			Available		
	records (including Near Miss incidents) and Corrective Action Preventive Action (CAPA) Reports				
5.13	Records in Training and experience in scaffolding for scaffolding supervisor or scaffolding gangs deployed		All Available	1	It is in Client scope
5.14	Periodic Inspection reports of T&P, Hand Tools, Power Tools, Welding M/Cs.		Not Available	0	no evidence found
5.15	Details of appreciations / penalties received in previous projects		Not Available	0	vendor found completed project without any penalty
5.16	Enclose photographs of Excavations executed (if any), Electrical installation, Stores / Storage area, Stacking Procedure Compliance		All Available	1	site photographs attached -refer to file photographs
				Overall Score	12

CAPABILITY EVALUATION REPORT OF CONTRACTOR						
(Civil, Mechanical, Electrical & Instrumentation)						
Sl. No	Evaluation Criteria	Guidance	Points	Scored Points	Audit Evidences- (Pl mention details in brief)	
6	**Statutory Requirements (Max Score – 4)**					
Complies to following statutory requirements, as applicable		**Applicable =0.5 ; Not Applicable =0**				
6.1	Availability of Service Tax / Sales tax		Available	0.5	GST available and attached refr to file pdf ketan	
6.2	Availability of Income Tax Clearance Certificate		Available	0.5	ITR for the year 2019-2020 attached refer to file ketan	

Sl. No	Evaluation Criteria	Guidance		Points	Scored Points	Audit Evidences
6.3	Availability of Trade License/Registration Certificate			Not Available	0	no evidence found
6.4	Provident fund registration (check for monthly remittance / wage register)			Available	0.5	PF certificates available and attached
6.5	Works Contract registration			Not Available	0	no evidence found
6.6	Employee's Insurance policy/Workmen Compensation policy			Available	0.5	certificate available and attached in file Ketan
6.7	Labour License	Check for number of workforce deployed vs actual labour obtained for		Available	0.5	available and attached
6.8	Minimum Wages			Not Available	0	taken care by client
Overall Score					2.5	

CAPABILITY EVALUATION REPORT OF CONTRACTOR					
(Civil, Mechanical, Electrical & Instrumentation)					
Sl. No	Evaluation Criteria	Guidance	Points	Scored Points	Audit Evidences- (Pl mention details in brief)
7	**General /Administration (Max Score – 5)**				
7.1	Adherence to Project Schedule (verify documents) can be covered earlier	Project plan versus achievement	Satisfactory	0.5	found satisfactory as per customer feedback letter. refer to section customer feedback above
7.2	Work Experience with Prestigious and major customers	Number of large customers in last three years . each customer marks= 0.25	3	0.5	Eldeco,TDI, Anshal Infrastructute and property pvt Ltd
7.3	Quality of Man Management at Site	Verify for conflicts by interviewing 5 workmen, verify	5	0.5	education, skill and experience found satisfactory

		persons deployed and education and experience, training and skills.			
7.4	See complaint records for past three years. (even 1 major non compliance its not met)	See complaint records for past three years.(even 1 major non compliance its not met)	Satisfactory	0.5	no evidence found for any non compliances
7.5	Capability of Handling Statutory agencies can be covered earlier	See reports sent to statutory authorities and comments received	Satisfactory	0.5	observed satisfactory at site
7.6	Mobilization Capability	(Workforce and T&P)	Satisfactory	0.5	observed satisfactory
7.7	Capability of handling local site conditions		Satisfactory	0.5	observed satisfactory
7.8	Accessibility to Contractor's Main office & Site office	Responsiveness of employees, approach etc	Satisfactory	0.5	found up to mark
7.9	Feedback from other customers	Customer feedback 5 Samples each sample carries 0.20 points	3	0.6	Checked customer feedback/site clearance report from following customer and found the same satisfactory 1 construction of boundary wall client name Eldeco Infrastructure and properties Ltd,Panipat,Haryana Job awarded date 1.4.2014 and completion date 25.7.2017 2 TDI ,Panipat ,Haryana-construction of road, contract awarded date -23.2.2017 and completion date 16.3.2018 3 TDI,Panipat,Haryana-construction of boundary wall, contract awarded date 1.12.2015 and completion date 30.3.2018
			Overall	4.6	

CAPABILITY EVALUATION REPORT OF CONTRACTOR
(Civil, Mechanical, Electrical & Instrumentation)

Sl. No	Evaluation Criteria	Guidance	Points	Scored Points
8	**FINANCIAL EVALUATION (Max Score -20)**			
8.1	**Sales Growth**		>20%	
	>20%	As per submitted evidences (Refer Last three years)	4.00	4
	>15%		3.00	
	>10 %		2.00	
	0 - 10 %		1.00	
	Below 0		0.00	
8.2	**PBT**		> 10%	
	> 10%	Growth over previous year (Average of last 3 years)	4.00	4
	> 8%		3.00	
	> 5%		2.00	
	0 - 4 %		1.00	
	Below 0		0.00	
8.3	**Growth of Net worth over previous years ([Net worth = Total Assets - Total Debts (Refer audited balance sheet)])**		> 5%	
	> 10%	High net worth relates to good financial strength and ultimately good credit rating for a company. Similarly a low or negative net worth will relate to a weaker financial strength and a lower credit rating, thus directly affecting the company's ability to raise funds from the market	4.00	2
	> 8%		3.00	
	> 5%		2.00	
	0 - 4 %		1.00	
	Below 0		0.00	
8.4	**Debt / Equity Ratio [Debt /Equity ratio= Total Long term liabilities (Debt) / Equity (Net Worth)]**		1:1	
	1:1	It is a measure of a company's ability to repay its obligations. If this ratio is increasing, company is being financed by creditors rather than from its own sources which may be a dangerous trend	4.00	4
	1:1 upto 1.5 : 1		3.00	
	1.6 upto 2.0 : 1		2.00	
	2.1 & above : 1		1.00	
8.5	**Inventory Turnover Ratio Cost of Goods Sold ÷ Average Inventory**		Above 4.0	
	Above 4.0	The inventory turnover ratio compares cost of goods sold with	2.00	2

	Between 4 and 2	average inventory for a period. This measures how many times average inventory is "turned" or sold during a period. It is important to have a high Inventory Turnover Ratio which shows the company does not overspend by buying too much inventory and wastes resources by storing non-salable inventory.	1.50	
	Between 2 and 1.25		1.00	
	Less than 1.25		0.00	
8.6	**Accounts Payable Turnover Ratio (Total Purchases ÷ Average Accounts Payable)**		Above 4.0	2
	Above 4.0	Total purchase is calculated by adding the ending inventory to the cost of goods sold and subtracting the beginning inventory. Average accounts payable is calculated by adding the beginning and ending accounts payable together and dividing by two. A higher ratio shows suppliers and creditors that the company pays its bills frequently and regularly	2.00	
	Between 4 and 2		1.50	
	Between 2 and 1.25		1.00	
	Less than 1.25		0.00	
Overall Score				18

CAPABILITY EVALUATION REPORT OF CONTRACTOR

Contact Details

Name	Designation	Mobile Number	Email id
Mr Rajbir singh	Quality Head	9354200008	ketanconstructioncompany@gmail.com
Mr Rajbir singh	Quality Manager	9354200008	ketanconstructioncompany@gmail.com
Chetak Chahal	Safety Head	9335700208	ketanconstructioncompany@gmail.com
Chetak Chahal	Safety Manager	9335700208	ketanconstructioncompany@gmail.com
Mr Rajbir singh	MD	9354200008	ketanconstructioncompany@gmail.com

PERFORMANCE RATING REPORT OF CONTRACTOR (SITE WORK)

Name & Address of Vendor	M/s	Name of the Auditor	Fulbodh chaudhary
		TPL EmpID	TPL91895
		SR Number	190540
		Date of Visit	08-Jul-2019
		Report No.	190540-080719-4
		Type of Service / Product	supply and modification of VFD PANEL
Details of Contractor's Representative interacted for evaluation: Here type the Name of the Representative		Name of the Site Referred	Same address as mentioned on the top left side
		WO No & Date of WO	2614&07-Nov-2018
		SBU Name	
Name	Mr	Brief Scope of Work carried out	
Cell No.		Supply and Modification of VFD Panels at Amanisha Nallah Project Site.	
Telephone No.			
Email		Project	JDA-AMANISHAH NALAH

Guidelines
1. Vendor's Performance is critical, impacting success of the project. Performance Evaluation shall be done with due diligence.
2. The scoring for performance rating needs to be carried out pragmatically and objectively.
3. Please do not evaluate based on single incident. Performance Evaluation shall be on overall performance.
4. Excellent or Not Satisfactory Ratings shall be substantiated with reasons separately by Project Manager.

Performance Levels

Scored Points	Performance Level	Action
91 - 100	Excellent	Appreciation Letter
81 - 90	Very Good	Appreciation Letter
71 - 80	Good	Feedback Letter
61 - 70	Satisfactory	Caution Letter
60 and below	Not Satisfactory	Delisting

Sl. No	Evaluation Criteria		Max Points	Scored Points
1	Progress & Mob	Adherence to Progress schedule	15	8
2		Timely Mobilisation of Essential Tools & Plants as per scope	10	4
3		Mobilization of workforce proportionate to work	10	3
4	IMS & Preservation	Quality of Works at Site	10	0
5		Compliance to Health, Safety and Environment Requirements	10	9
6		Handling, Storage & Preservation	6	2
7	Stat. Finance & Recording	Statutory Compliances by Contractor	6	0.5
8		Financial Capability during Execution	6	3
9		Periodic Reconciliation of Free-issue Materials	6	0
10	Retention & Responses	Response to (TPL / Client) Instructions	6	4
11		Retention capability of workforce during execution	5	0
12	Documentation & HK	Documents Control	5	3
13		Housekeeping	5	5

		Total	100	41.5

Overall Observation: 1 Vendor response to evaluation was prompt and positive 2 Vendor delivery compliance observed not satisfactory as there is delay in meeting customer requirement 3 No evidence found for customer order scheduling and subsequent production planning and control

Strength: 1 compliance to instruments calibration found satisfactory 2 Housekeeping observed up to mark on shop floor 3 Skill matrix and training plan observed good

Opportunity of Improvement: 1 Vendor was unable to recall their supplier purchase orders used in production of subject materi to customer. Advise to maintain batch tractability to track supplier PO. 2 Vendor found not using any identification tags and inspection status tags for in process material although they have got ISO 9001 certification 3 customer complain register was found incomplete. complain found not closed 4 no evidence found for customer complain 7 step problem solving to ensure corrective and preventive action

Initiated by: Name, Signature & Date	Fulbodh chaudhary
Reviewed by: Name, Signature & Date	Venu Gopal B

*For Cat-2 and Cat-3 Items
Format No – 06.01.02 F-02 R2

Sl. No.	Sub-criteria	Checkpoint	Sub-Criteria no	Guidance Note	User Entry	Scored Points	Audit Evidence (Brief description of the evidences)
1	Adherence to Progress schedule	Adherence to Progress schedule > 100% 91 % to 99 % 81 % to 90 % 71 % to 80 % <70%	1.1.A.1	Check the Project schedule & Progress Report	71 % to 80 %	8	Contractor has complied to supply of materia as per schedule but for modification and commissioning n evidence found. PO copy and bill copy has been attached in link submission. Schedule compliance sheet has been attache
2	Timely Mobilisation of Essential Tools & Plants as per scope	Adequate Mobilisation of Tools & -Plants in good condition * Mobilised fully * Mobilised partially * Failed to mobilise	2.1.A.1	Check the Project Plan & Deployment schedule and verify if mobilisation completely done with suitable equipment and free from damages and fit for use. (also Gate pass, DC or LR could be verified where reqd)	Mobilised partially	2	no evdience foun but as per the Invoice the work was completedo 30.11.2018
				Mobilisation < 70% / inadequate and few T&P noted in damaged condition			
				No evidence of Mobilsation or Mobilisation Plan or most of the Tool & Plants mobilised are in damaged condition			

		Timely mobilisation of necessary Tools & Plants (3 Points) * Mobilisation in time * Missed the schedule * Failed	2.2.A.1	Mobilisation done as per the Project Plan & Deployment Schedule timelines (as per work order and subsequent amendments if any)	Missed the schedule	2	no evidence found, but as per the Invoice the work was completed on 30.11.2018
				Delayed Mobilisation but done			
				Mobilisation not initiated			
		Working Condition of Tools & Plants- * Availability of valid test / fitness certificates from authorised agencies (2 Points) * Carrying out Preventive & Regular Maintenance (2 Points)	2.3.A.1	Review certificates from Govt approved bodies/lab for at least 4 Types of Tool & Plants e.g. cranes, welding machines, INC, lifting slings winches etc	0	0	no evidence found
			2.3.A.1	Review Maintenance Records/ Log Book/ T&P Register	0	0	no evidence foun
			2.3.B.1	review records of at least 2 equipment	0	0	no evidence found
3	Mobilization of workforce proportionate to work	Timely mobilisation of required workforce w.r.t. agreed deployment schedule	3.1.A.1	All Workforce mobilised as per Project Plan/ Deployment Schedule	Delayed deployment	3	no evidence foun but as per the Invoice the work was completed o 30.11.2018
		Inadequate mobilisation of workforce w.r.t. work front not affecting the project schedule		Only 60 % mobilised as per plan/schedule, but has not affected the project schedule			
		Delayed deployment of Insufficient workforce w.r.t. work front affecting the project schedule		Less than 60 % of the workforce not mobilsed as per plan/schedule, but has affected the project schedule			
		Not Mobilised/ Demobilised entire workforce before completion of job		Workforce not mobilised at all or demobilised before completion			
4	Quality of Works at Site	Compliance with FQP * > 95% * 80% -95% * < 80%	4.1.A.1	Verify all parameters & stages mentioned in the FQP also check the Fortnightly Field Quality Report (FFQR) from the FQE, if > 95% compliant 80% -95% compliant < 80% compliant	< 80% - 0	0	no evidence foun for work at site

		Resolution of NCRs and completion of Reworks within stipulated time * 100% * Less than 100%	4.2.A.1	Take list of NCR's, from FQE and check for timely completion of rework / resolution of NCR's for suitable sample size, Iso check the FQP/FIN where reqd, if 100% resolved / completed on time or less 100 % resolved/ ompleted on time	Less than 100%	0	no evidence foun for FQE/NCR
		Timely preparation & submission of Protocols * 100% * 80% -99% * <80%	4.3.A.1	Check if Joint Inspection Reports (Protocols) are available at appropriate stage with the vendor/ FQE and is submitted by the vendor as defined in FQP. If 100 % Protocols are submitted on time if > 80% and < 99% submitted on time if < 80% protocols submitted on time	< 80%	0	FQP/FQE not found
		Number of stages cleared from inspection in first attempt * > 95% * 90% -95% * 85% - 90% * Less than 85%	4.4.A.1	Check the Job Offering Register/ Request For Inspection (RFI) for various stages, verify stages as mentioned in the FQP . If > 95% of Stages cleared in 1st attempt if > 90% and < 94% of Stages cleared in 1st attempt if > 85% and < 89% of Stages cleared in 1st attempt if < 84% of Stages cleared in 1st attempt	Less than 85%	0	no evidence foun for stage inspection
5	Compliance to Health, Safety and Environment Requirements	Use of PPE & other safety gadgets 100%, 95 - 99%, Less than 95%	5.1.A.1	Observe whether the workforce use the right PPE & other safety gadgets	100%	3	taken care by client
		Compliance to Safe working conditions * 100%, * Less than 100%	5.2.A.1	Check SCR, SAR, SRM and look for incidents/ near miss or also Incident Register /Near Miss report or Medical History Register etc.	100%	4	taken care by client
		Providing Safe Drinking water & Sanitary facilities at labour camp	5.3.A.1	5.3 a). Check the frequency of checking the Water Quality and check reports	Yes	1	taken care by client
			5.3.B.1	5.3 b). Check for availability of Washroom /Toilets are they are maintained/ hygiene conditions	Yes	1	taken care by client

		Waste Disposal / Eco-friendly Practices	5.4.A.1	Check for availability of adequate arrangements (separate bins) for waste management to be verified like red bin, Yellow bin & green Bin Hazardous waste & Non Hazardous waste identified	No	0	no evidence foun
6	Handling, Storage & Preservation	Handling of materials / equipment as per handling instructions / standard practices	6.1.A.1	Availability of Cranes, Hydra, Winches, Hydraulic Jacks, Chain Pulley Blocks, Wire Ropes	0	0	no evidence found
			6.1.B.1	Equipment in working condition	No	0	no evidence found
			6.1.C.1	Personnel are trained in using the respective equipment	No	0	no evdience found
			6.1.D.1	Any Instructions available e.g. SOP/WI/Manufacturer Manual	No	0	no evidence found
		Storage & Preservation Maintains covered shed, follows First In First Out (FIFO) for shelf life items, follows manufacturer's instructions and maintains Identification and Inspection / test status.	6.2.A.1	Shed available	Yes	1	taken care by client
			6.2.B.1	FIFO followed for items with shelf life	Yes	1	FIFO Nnot applicable
			6.2.C.1	Items stored have Identification & Inspection status	No	0	no evidence found

Sl. No.	Sub-criteria	Checkpoint	Sub-Criteria no	Guidance Note	User Entry	Scored Points	Audit Evidence (Brief description of the evidences)

7	Statutory Compliances by Contractor	Maintaining PF/ESI /WC/ All Insurances in contractors scope	7.1.A.1	Check from the Site HR ADMN for availability of (i) PF Registration (ii) ESI scheme provision (iii) Works Contracts (iv) Workmen Insurance/compensation policy	0	0	no evidence found
		Maintaining validity of applicable work-specific licenses (IBR / Electrical works / any other specific license)	7.2.A.1	Check from the Site HR ADMN for availability of-(i) FORM IV -Application for Licence under [Rule 21(1)], (ii) FORM V- Form of Certificate by Principal Employer [See rule 21(2)], (iii) Copy of Work Order, (iv) Challan Fee & Security Deposit	1	0.5	Evaluation was done at vendor office and hence HR could not be contacted. copy of work order
		If Labour license is valid or Evidence of Application for Renewal available for Invalid license	7.3.A.1	If Labour license is valid or Evidence of Application for Renewal available for Invalid license	No	0	no evidence found
8	Financial Capability during Execution	Ability to manage works without Financial constraints (a) workforce (b) equipment (c) supplies/services	8.1.A.1	Check if any activity was halted due to nonavailbility of (a) workforce	No	1	no evidence found but as per the Invoice the work was completed on 30.11.2018
			8.1.B.1	Check if any activity was halted due to nonavailbility of equipment	No	1	no evidence found but as per the Invoice the work was completed on 30.11.2018
			8.1.C.1	Check if any activity was halted due to nonavailbility of supplies/services from manufacturer/sub-contratcors (also infere information from Criteria 2 & 3)	No	1	no evidence found but as per the Invoice the work was completed on 30.11.2018
		Payment of wages on time to their workforce	8.2.A1	Check at least 3 employees in the Wage Register	No	0	no evidence found
9	Periodic Reconciliation of Free-issue Materials	Reconciliation and monthly updating of records	9.1.A.1	Check last 3 cases of Free Issue Material- and timely RECO is done - check the Vendor Stock Register/ Debit Notes, Material Receipt and Issue Voucher (MRIV)- From Site Stores	0	0	no evidence found
			9.2.A.1	If agreed norms/Cutting plan/Rolling margin available	No	0	no evidence found
			9.2.B.1	Check 2-3 cases whether within norms and enter the count	No	0	no evidence found

Sl. No.	Sub-criteria	Checkpoint	Sub-Criteria no	Guidance Note	User Entry	Scored Points	Audit Evidence (Brief description of the evidences)
10	Response to TPL / Client Instructions	Response to TPL / Client Instructions	10.1.A.1	Take Data from CE or SS or RCM (Speedy is before time and Prompt is Just in Time)	Prompt	3	no evidence found for site work
		Response to change in scope	10.2.A.1		Slow	1	no evidence found
11	Retention capability of workforce during execution	Has full control over retention of the work force	11.1.A.1	More than 90% of the workforce continue as being employed	No Control	0	no evidence found for attendance sheet
		Attrition of workforce not affecting the project schedule		Around 30% have resigned or left but the project schedule is not affected			
		No control on retention of workforce		More than 50% have resigned or left and has significantly affected the project schedule			

Sl. No.	Sub-criteria	Checkpoint	Sub-Criteria no	Guidance Note	User Entry	Scored Points	Audit Evidence (Brief description of the evidences)
12	Documents Control	Ensures availability of Latest approved Drawings, FQP, Procedures, Formats at work place and removal of superseded documents	12.1.A.1	Check availability of current Drawings, FQP, Procedures, Formats	0	0	no evidence found
			12.1.B.1	Check if the above are well maintained and are legible	Yes	0.5	AS per quality manual and iso certification standard
			12.1.C.1	whether superseeded documents removed from place of work	Yes	0.5	during manufacturing ensured as per quality manual clauses
		Ensures valid Calibration certificates, Inspection/Testing reports & Protocols	12.2.A.1	Inspection Measuring & Testing equipment are calibrated	Yes	0.5	have valid calibration certificate

			12.2.A.2	Records available & traceable to National or International Standards	Yes	0.5	found NABL accredited
			12.2.B.1	Previous testing/inspection reports are available	Yes	0.5	found available as per list attached
			12.2.B.2	Available approved by authorised personnel	Yes	0.5	calibration certificate available
		Timely submission of MTCs and or As-built drawings	12.3.A.1	Built drawings submitted on timely basis	No	0	no evidence found
			12.3.A.2	MTC submitted on timely basis	No	0	no evidence found
13	Housekeeping	Well organised Satisfactorily organised Not Satisfactory	13.1.A.1	Check visually if (i) the work place /site are well lit (ii) storage layout organsied (iii) items identified and stored in designated areas (iv) gangway for easy movement (v) safe storage practices followed	5	5	All found complied during manufacturing

Abbreviations	
CE	Construction Engineer (Concerned)
CFTAR	CFT Audit Report
FIN	Field Inspection Note
FQE	Field Quality Engineer (Concerned)
FQP	Field Quality Plan
NCR	Non-Conformity Report
SAR	Safety Compliance Report
SAR	Safety Audit Report
SCM	Safety Committee Meeting
SE	Safety Engineer (Site)
SRM	Safety Review Meeting
SS	Site Stores-in-charge

www.ingramcontent.com/pod-product-compliance
Lightning Source LLC
Chambersburg PA
CBHW080438220526
45465CB00009B/3342